Livestock on Dry Land Farms

With Information on Keeping Horses, Cattle and Sheep on the Dry Farm

By

Thomas Shaw

British Library Cataloguing-in-Publication Data
A catalogue record for this book is available from the
British Library

Farming

Agriculture, also called farming or husbandry, is the cultivation of animals, plants, or fungi for fibre, bio-fuel, drugs and other products used to sustain and enhance human life. Agriculture was the key development in the rise of sedentary human civilization, whereby farming of domesticated species created food surpluses that nurtured the development of civilization. It is hence, of extraordinary importance for the development of society, as we know it today. The word *agriculture* is a late Middle English adaptation of Latin *agricultūra*, from *ager*, 'field', and *cultūra*, 'cultivation' or 'growing'. The history of agriculture dates back thousands of years, and its development has been driven and defined by vastly different climates, cultures, and technologies. However all farming generally relies on techniques to expand and maintain the lands that are suitable for raising domesticated species. For plants, this usually requires some form of irrigation, although there are methods of dryland farming. Livestock are raised in a combination of grassland-based and landless systems, in an industry that covers almost one-third of the world's ice- and water-free area.

Agricultural practices such as irrigation, crop rotation, fertilizers, pesticides and the domestication of livestock were developed long ago, but have made great progress in the past century. The history of agriculture has played a major role in human history, as agricultural progress has been a crucial factor in worldwide socio-economic change. Division of labour in agricultural societies made (now) commonplace specializations, rarely seen in hunter-gatherer cultures, which allowed the growth of towns and cities, and the complex societies we call civilizations. When farmers became capable of producing food beyond the needs of their own families, others in their society were freed to devote themselves to projects other than food acquisition. Historians and anthropologists have long argued that the development of agriculture made civilization possible.

In the developed world, industrial agriculture based on large-scale monoculture has become the dominant system of modern farming, although there is growing support for sustainable agriculture, including permaculture and organic agriculture. Until the Industrial Revolution, the vast majority of the human population laboured in agriculture. Pre-industrial agriculture was typically for self-sustenance, in which farmers raised most of their crops for their own consumption, instead of cash crops for trade. A remarkable shift in agricultural practices has occurred over the past two centuries however, in response to new technologies, and the development of world markets. This also has led to technological improvements in agricultural techniques, such as the Haber-Bosch method for synthesizing ammonium nitrate which made the traditional practice of recycling nutrients with crop rotation and animal manure less important.

Modern agronomy, plant breeding, agrochemicals such as pesticides and fertilizers, and technological improvements have sharply increased yields from cultivation, but at the same time have caused widespread ecological damage and negative human health effects. Selective breeding and modern practices in animal husbandry have similarly increased the output of meat, but have raised concerns about animal welfare and the health effects of the antibiotics, growth hormones, and other chemicals commonly used in industrial meat production. Genetically Modified Organisms are an increasing component of agriculture today, although they are banned in several countries. Another controversial issue is 'water management'; an increasingly global issue fostering debate. Significant degradation of land and water resources, including the depletion of aquifers, has been observed in recent decades, and the effects of global warming on agriculture and of agriculture on global warming are still not fully understood.

The agricultural world of today is at a cross roads. Over one third of the worlds workers are employed in agriculture, second only to the services sector, but its future is uncertain. A constantly growing world population is necessitating more and more land being utilised for growth of food stuffs, but also the burgeoning mechanised methods of food cultivation and harvesting means that many farming jobs are becoming redundant. Quite how the sector will respond to these challenges remains to be seen.

LIVE STOCK ON DRY FARMS

When the homesteader locates on the dry farm, his efforts are usually concentrated, and properly so, on the production of grain, but he makes a serious mistake if he entirely neglects the keeping of live stock, for the presence of the cow and the brood-sow are about as essential to the farm home in dry areas as the presence of the breaking plow. It is true, nevertheless, that live stock on the dry farm should not be introduced with undue haste, for at the outset the furnishing of food for winter may prove a costly problem in seasons that are unusually dry.

That the production of grain for sale should be the principal object of the dry farmer during the first years of his farming is undoubtedly true, but in time more or less of live stock should be grown upon his farm. This should be done to the extent of using practically all the coarse grains that he will grow and also the hay and straw produced as well as the pasture areas that are accessible on the farm or on the unoccupied lands that may be adjacent thereto.

The lament that the tillage of the arable areas of the open range is going to destroy the live stock industry in dry areas is not well founded. Even on the arable farm devoted largely to the growing of grain, more live stock can be kept in addition than were formerly kept on a similar area. This results from the greatly increased production that follows the proper tillage of the soil in fodder and also in pasture. It would seem safe to say that the food nutrients in the straw grown on an acre of well tilled land in dry areas will be more than the food nutrients from an acre of the same before the land has been broken. The food nutrients produced by an acre

1

of well planned pasture from grasses grown under cultivation, should be from two to three times as much as from grasses produced from the native prairie. The production of live stock on the arable farms will therefore, in time, greatly increase the production of beef and mutton, to say nothing of the production of pork and poultry, which was impossible under old-time range conditions.

Why live stock should be kept.—Live stock should be kept on the arable form for the following reasons among others that may be given: (1) to prevent waste on the farm; (2) to prevent waste on the range; (3) to increase diversity in production; (4) to maintain fertility in the land; (5) to furnish food for the home, and (6) to increase the revenues from the farm.

In the absence of live stock on the arable farm other than the work horses that till the land, serious waste is unavoidable; (1) there will be more or less waste in the uneaten grasses of untilled portions; (2) in the straw, much or all of which will probably be burned; (3) in the grain heads that are lost amid the stubbles because ungleaned; (4) in the uneaten food that grows up amid the stubbles, and (5) in the unconsumed grain that is unavoidably wasted where threshing is done in the open air. Such waste is unavoidable in areas where live stock is not maintained.

In newly settled areas, there are usually more or less range pastures contiguous to the individual farms. This may be entirely wasted in the absence of live stock to consume it. In some instances these pastures are so ample as to justify the homesteader in making the growing of live stock the dominant feature of his work, until the adjacent lands are taken up as homesteads.

The growing of live stock encourages diversity in production. It encourages the growth of forage and root crops. While they are being grown the land is being prepared for the successful growth of grain crops the following year. The larger, therefore, that the area of such crops is, up to the limit of the ability of the farmer to properly care for them without undue outlay for hired help, the larger should be his profits from the ground thus tilled, as, to the extent that he grows these crops, he avoids the necessity for the cropless bare-fallow.

The introduction of live stock makes it possible to maintain virtually undiminished production in the land. The waning fertility in lands where such production has been long deferred, as in some parts of the Dakotas, California and other states, should serve as a warning. Some lands, and especially the volcanic ash soils of the west, may stand continued cropping for many years, but ultimately they must fail. Beyond all question, the fertilizing material produced by live stock will add greatly to production in all its various lines.

Live stock may be made to furnish a large part of the living of the farmer, and with but little cost. The cow, for instance, can turn the free grasses of the prairie into the best food that man may get. The brood-sow and her progeny will manufacture the same grasses with waste from the grain fields into meat for the winter. Fowls with a moderate grain supplement will turn grasshoppers and other insects into valuable food. Such live stock, therefore, should be introduced the first season where there is a family on the ranch, but, of course, in very limited numbers.

At the very outset, therefore, the revenues of the farm may be increased by reducing the outlay for food. The farmer with a family who fails to try to grow a large part of his living on his own farm is not true to himself. As time goes on the live stock will become a source of considerable revenue, although for several years it is not likely to become the chief source of revenue on the arable farm.

The kinds of live stock to grow.—Because of the varying conditions, the discussion of this question is not easy. These are such as relate to the character of the production, the location and its surroundings and the predilections of the individual. It will be manifest that in some areas the production of grasses will be relatively easy, in others relatively difficult. In one location pasture land may be cheap and relatively plentiful. In other instances it may not be possible to secure it outside of the home. One man may succeed best in handling dairy stock, and another will succeed best with sheep. All these and other factors must be considered.

Among the determining factors are the following: (1) the climate; (2) the precipitation, and (3) the market. The climate has an important bearing on the shelter called for and also on the production. Fortunately, in' much of the dry area of the west, shelter is not so much needed as in corresponding latitudes in the east, but some shelter is called for, and more for some classes of live stock than for others, and all shelter is more or less costly. The precipitation has an important bearing on the production. For instance, in the upper Flathead valley, 15 inches of rainfall produces more pasture and more succulent than a similar rainfall in the same latitude west of the mountains, hence dairying may be more readily conducted in the former than in the latter. The market demands alone may determine the character of the live stock that ought to be chiefly grown, and the facilities for marketing should also be carefully considered. The farmer with a good local market near at hand has a great advantage over the one whose market is not local and distant.

All farmers, whether married or single, must have horses. That question does not admit of discussion. Beyond that much will depend on the presence or absence of a family in the home. Where it is present, it

4

is, in a sense, imperative to have enough live stock to supply the needs of the family. This means that the farmers must, in a sense, have some cows, swine and poultry, and it will be all the more to their interest to have some sheep, especially after the farm is fenced. The farmer thus equipped, who at the same time grows his own vegetables and small fruits, has but little additional outlay for his living. To furnish this he does not need to maintain a large number of any of these classes of animals.

Under some conditions it may be wise to extend the growing of live stock so as to make it the dominant industry, even at an early day in the work of the homesteader, as, for instance, where free or very cheap pastures are easily accessible. Such extension may apply to horses, cattle or sheep, according to the conditions that may be present.

Stocking the dry farm.—As a rule, the live stock on a dry farm should be introduced very gradually. For this several reasons may be given. First, there has not been time to make suitable preparation to care for them in a large way. The outbuildings are not ready. Fences have not been built. It may be that winter water supplies have not been secured. There has not been time to make sure of winter supplies of food, as the bulk of the ground broken is usually wanted for grain. The introduction of much stock by purchase is costly. It is much better, as a rule, for the dry land farmer whose operations must be confined to his own farm, to begin with a small amount, and to grow much or all of the subsequent increase. When increase is made in this way, all the operations of the farm may be kept in due balance. Such increase is thus obtained at a minimum of cost. The experience while making it has been obtained under the attendant conditions, and is, therefore, doubly valuable. The protection wanted may also be furnished at a minimum cost.

The amount of live stock that a dry land farm will sustain cannot be stated. It cannot be even approximated, owing to the very great difference in the conditions. It will be at once manifest that where the rainfall is 18 inches, much more live stock can be kept than where it is but 12, or even 15 inches. It will also be apparent that where bulky foods may be readily grown, as corn and the sorghums, more live stock relatively may be kept than in areas where these do not succeed well because of low temperatures. It would seem safe to say, that the amount of live stock that a dry farm will sustain will increase with increase in the precipitation. The relative number that such a farm will sustain is not high, not so high as in humid areas. Such farming where the farmer has no access to outside grazing lands is usually a mixed farming proposition in which the growing of grain for sale will probably be a dominant factor for many years to come. It would be unwise, therefore, for one situated thus to make the growing of live stock a dominant factor in his work at the outset, but this may be done by the farmer who has access to pastures that are cheap or free.

Great care should be exercised not to overstock the dry farm. Under the most favored conditions, such a mistake is very costly, as it forces the sale of the animals, whether in good condition or lean, and at such a time it is almost certain that they will be lean. When a very dry season comes, and it may come at any time, there may be a serious shortage in both pasture and fodder, hence some reserve kept over from a season more bountiful may be a wise provision. While the hazard mentioned may occur, it does not furnish a fitting excuse for the exclusion of live stock from any farm.

Growing horses.—On the dry farm horses will always be necessary in order to do a large part of the work. This statement does not mean that other kinds of power, as steam and gasoline, may not be extensively used, especially in breaking up the stubborn soils of the prairie. The horse will always •be in evidence not only on the dry farm, but on all farms.

Usually from three to four horses are called for when breaking up the stubborn soils of the prairie. There would seem to be no good reasons why two at least should not be brood mares. These may produce foals while aiding in doing the work of the farm, providing they are carefully worked. As much work should not be exacted of them as if they were not suckling foals, but they will still do a large amount of work and also rear foals if well fed. It will, furthermore, be a decided advantage if such foals come in the autumn, for then the dams can suckle them at that season of the year when they are not worked, as they are in the summer. Experience has shown that brood mares worked in moderation will rear foals more surely and in better form than those that are not worked at all. The young horses, if of the draught types, may be made to aid in the light work of the farm when from 2 to 3 years of age, and when they have reached the latter age, they will sell readily for a good price.

Horses reared in the dry country call for but little shelter at any season of the year. When allowed liberty, they can secure food where other classes of live stock would not be able to do so, as when, for instance, the ground is covered with snow. The habit of pawing to remove the snow makes it possible for them to live and flourish where other classes of live stock would not survive under similar conditions. In the winter they will utilize such foods as straw to better advantage than most other live stock. Relatively, therefore, they may be grown cheaply.

The return is also large, for the numbers kept. Where the farmer devotes his attention mainly to the growing of this class of stock, he does not need to have many of them on the place at one time, hence there is but little hazard of loss in a season of drought. Where range of rough pastures is accessible, in the larger portion of the dry country, horses will come through the winter in good form without the necessity of very much supplemental food.

The supplemental foods for the feeding of growing foals include alfalfa, fodder corn fed in the bundle, and straw, in northern areas. For idle work horses, straw will suffice for much of the winter. In southern areas, alfalfa, Milo maize, Kaffir corn and sorghum will best answer the purpose. Milo maize is also much esteemed for feeding work horses in these areas. It is usually fed to them as grain food in the head, which not only obviates the necessity for threshing the grain, but it also insures a more complete digestion of the naturally hard grain, since it is more thoroughly masticated.

Growing dairy cattle.—That the dry farm is not nearly so well adapted for dairying as the irrigated farm cannot be gainsaid. The creamery, therefore, will not probably be much in evidence where dry farming is followed for many years subsequent to the settlement of the land. But this does not mean that home dairying may not be practised in a moderate way, even at the outset of the farming. It may even be wise in some instances for a farmer to give considerable attention to dairying at the outset, where free pasture is plentiful, but when the stock must be confined to the limits of the arable farm, the number of cows kept should not usually be large.

Where the cows are to be kept within the limits of the farm, the grazing problem is more difficult than the furnishing of winter foods. The pastures that may be grown are discussed elsewhere (see p. 355). These, however, may be supplemented by such soiling foods as alfalfa, corn and field roots in the north, and by alfalfa, the sorghums, Milo maize and stock melons in the south.

The winter food in the north may be made up almost entirely of alfalfa hay and corn fodder fed in the bundle, and supplemented by a very small amount of rye, barley or speltz fed in the ground form. The ration may also consist of alfalfa and grain cut underripe and fed unthreshed. A mixture of Canada field peas and beardless barley grown together is excellent. This also is true of millet grown in rows and cultivated. In southern areas the winter food may consist mainly of alfalfa hay, sorghum and Milo maize fodder, and a supplement of Milo or sorghum meal, or of ground speltz. The Milo maize may be fed as seed and heads ground together.

The cows should drop their calves in the fall rather than the spring, as the calves can be fed with more care and success in the winter season. The cows will then be dry in those months of the late summer when the pastures are dry. When thus managed, the milk flow will be better sustained, and the lactation period more or less prolonged.

As such cows will, of course, be hand-milked, the skim milk and the buttermilk should be fed to calves and swine. The butter or the cheese product, as the case may be, will usually find its way to private customers, as it will be too restricted in quantity for a wholesale market.

Growing beef cattle.—In dry areas the field for growing beef is probably wider than that for growing

9

dairy products, owing to the fact that it may be grown largely, in many instances, on rugged and broken pastures in proximity to the arable farm or forming a part of it. During the milk period, beef will be produced by methods that are radically different. On the strictly arable farm, the calves will be hand-fed, while on the arable and rugged farm, they will be suckled by their dams.

When reared by hand, the calves should be progeny of dual cows, and the aim should be to have them come in the fall. If the progeny of dairy cows, they should be sired by a beef bull. They should be reared essentially on skim milk and adjuncts after the age of two weeks. They should have good grazing, as rye or rape, sown especially for them, and they should, as a rule, be put on the market at the age of not more than 18 months. While taking milk, and subsequently, such meals as bran, ground oats, barley or Milo maize should be fed to them freely, also a nice quality of alfalfa hay. During the second winter they should be fed on such fodders as alfalfa, corn and sorghum, and should get a few pounds daily of such meal as barley, speltz or Milo maize. The aim should be to force growth with a prudent haste and thus shorten the period of pasturing and effect a substantial saving in the food of maintenance.

When reared on the dams during the milk period, it may not always be the best plan to have the calves come in the fall, as when they come in the spring the cows may oftentimes graze much of the winter on the rugged pastures. Provision should be made for saving such pasture by keeping the stock from grazing on that portion in summer. The first winter the calves should be given a moderate amount of grain along with the fodders named above. The second summer they will be on the pastures without grain. The second winter they may be fed· similarly to the hand-reared calves when preparing them for market, and they ought to be sold when two years old. Some western farmers head their wheat and feed it thus to cattle that are being fattened. The wisdom of doing so is to be questioned. They should then weigh about 1,200 to 1,300 pounds, while those hand-fed would weigh about 1,000 pounds.

Such animals should find a ready market wherever high-class meat is wanted. Grown thus it will be high-class meat, and should command the highest price. When growing such meat, it should never be allowed to become lean.

Sheep on the dry farm.—When sheep are reared on the dry farm the number so reared should not be very large. The production on the same when mixed in character would not justify the maintenance of numbers so large relatively as in humid areas. In the latter, sown pastures may be grown in succession through all the season, but only to a limited extent in the former. The numbers, however, should be enough to consume all the forage that would otherwise go to waste, as, for instance, grazing in the lanes and amid the stubbles in the fields and on summer-fallow land. On every farm the pasture from such a source is ample to sustain a small flock of sheep, which may be thus grazed virtually in most localities for three-fourths of the year without cost. The benefit thus rendered will be marked in the destruction of weeds, and in distributing more or less fertilizer over the land.

When rough pasture areas are contiguous to or form a part of the dry arable farm, sheep can be grazed on them about in the same way as cattle are grazed as described above. The breeding portion of the flock may be wintered on food grown on the arable portion of the farm. The part of the flock to be disposed of may be finished on pastures grown for the purpose. These may comprise Dwarf Essex rape grown in rows and culti-

11

vated as a rule, some dwarfish kind of corn or peas and white hulless barley. These will be grazed where they grew, and, if necessary, they may be supplemented by more grain grown on the farm, as, for instance, barley in the north and Milo maize threshed or in the head in the south.

SHEEP ON RANGE PASTURES, MALHEUR COUNTY, OREGON.
Courtesy Great Northern Railway Co.

When simply fattened on the farm the supplies for fattening will be purchased and they will be finished on foods such as have been referred to in the preceding paragraph. In the San Luis valley of Colorado the fattening of sheep mainly on peas has grown into a large industry. There are no reasons for concluding that this industry may not be extended to many other areas.

The opportunity to fatten sheep on Squaw corn or some other small variety grown for the purpose in the dry area, is thus virtually without limit. The growing of the corn stores moisture in the soil for the next crop, and feeding it off thus furnishes readily available fertility. Such grazing is not well adapted to areas in which considerable rain or snow falls in the autumn months.

The market for the surplus of the small flock should be found chiefly on the farm itself. It should furnish an important source from which the home supply of meat may be obtained, especially in the winter season. When finished in a wholesale way, as on crops that are grazed, they will be in condition to meet the needs of any market that may be accessible to them.

Swine on the dry farm.—The place for swine on the dry farm will always be one of considerable importance. From this source, more than any other, must come the supply of farm meats. To purchase meat for the home on the dry farm that may be all grown upon it would be a grave mistake, owing to the fact that pork may be slaughtered at almost any age should necessity call for it. Complications which sometimes arise with other animals from over-stocking may be prevented. Where a few cows are kept, the milk will aid greatly in giving the swine a good start in growth at an early age.

But even in the absence of cows, swine may be grown with much profit. The young pigs should remain on the dams until they become self-weaned, and the farmer should be content, as a rule, with but one litter from each dam in one year. With the aid of skim milk, two litters may be grown save where the winters are quite cold. The dry land farmer is not in a position to grow swine as cheaply or as numerously as the farmer who grows the same in irrigated areas. In summer alfalfa is the basic pasture for swine in the larger portion of the dry country. But other pasture, as Dwarf Essex rape, beardless and hulless barley, is good. In the south the sorghums may be made to add to the grazing; northward swine may be finished in the fields, as on peas or some small kind of corn. In the south peanuts and also corn furnish good fattening foods. If Milo maize is fed it should not be fed alone, as the swine do not thrive when fed thus.

On the dry farm the needs of the family must first be supplied. If there is a surplus beyond this, it will

SWINE GROWN ON DRY LAND FARM, NEAR HOBSON, MONTANA
Courtesy Great Northern Railway Co.

usually find a market without the need for shipping it. This is one of the few food commodities that will never glut the market, but it may be necessary sometimes, as when large numbers are grown and fattened on products in the field, to ship the animals away by rail.

Poultry on the dry farm.—Poultry should be kept on every farm where there is a family. The climatic conditions for growing it are, in nearly all localities, very excellent, because of the dryness of the air and its temperate character. The relatively small amount of grain food called for makes the growing of fowls a safe and profitable business, where it is wisely conducted. Even where grain could only be grown once in two years on the same land, the business should be a safe and profitable one.

All the grain foods grown in dry areas may be fed with profit to poultry if suitably blended. Of these none is better or even quite so good as wheat. Hulless barley is excellent. On southern areas the seed from Milo maize and the sorghums will serve an excellent end. These are frequently fed by suspending the heads on a stretched wire and allowing the fowls to help themselves, thus furnishing them with needed exercise, especially when they have to reach high for the food. The varieties with a bent down head are most easily suspended thus. The green food may come in the form of alfalfa, rape and field roots.

The demand for these products will always continue good, as they are staples that will always be wanted, but the product can be readily transported and at moderate cost for the value. In the home the product is indispensable, and the food which it furnishes is wholesome in character. There is probably no other class of live stock that can be grown on the dry farm that will yield a larger profit on the investment or that is more easily conducted.

The size for the dry farm.—The size for the dry farm should be determined by such considerations as: (1) the amount of the precipitation; (2) the character of the soil; (3) the capacity of the individual, and (4) the style of the farming.

It is, of course, impossible to determine the relative influence which these considerations should exert, but the first consideration is certainly one of much importance. It would seem safe to say that the area to be farmed should increase as the precipitation decreases, for the less the amount of precipitation, the fewer is the number of the crops that can be grown in a given term of years. It is very evident that the farmer who can grow but one crop in two years should have more land to till than the farmer who may expect to get a crop practically every year. It was this consideration that led to the granting of homesteads in certain areas of 320 acres instead of 160 acres, the usual size for such farms.

The tillage of a soil that is naturally friable and that holds moisture readily does not call for so much labor as the tillage of a heavy soil, hence the farmer whose soil is of the first class can till a much larger area than the farmer whose soil is of the second class, and with no greater expenditure of labor. When tilling the heavy soil, the compensation may come from larger yields, at least in some instances. On general principles, therefore, the lighter the soil, the larger should be the area that is capable of being farmed.

When determining the size of the farm, much depends on the capacity of the farmer. It is certainly safe to assume that the less the capacity of the farmer, the smaller should be the amount of the land which he tills. One farmer with large capacity may handle fairly well a whole section, or even more than a section, whereas another farmer may not have capacity to handle well a quarter section. Something depends on the farmer's family. A farmer whose family is sufficiently grown to enable him to do his work without hiring should succeed better on a farm large enough to utilize all the labor than on one of less size. As wages are at the present time, the dry land farmer should sedulously aim to avoid hiring to the greatest extent possible, and to accomplish this end when investing he should gauge accordingly the size of the farm that he can till.

There is another class of farmers whose work is in a sense speculative. They live in the cities. They usually own large areas and farm them in a speculative way. These men are wholly dependent on hired labor, hence in order to get a remunerative return they must of necessity farm large areas and in a wholesale way. Such farming may be successful as long as the land is new and clean, but in all states the story of such farming is the same. Within a few years the land usually becomes very foul with weeds and the crops become so unproductive that tillage operations result in loss, but substantial profits may be realized, nevertheless, from the advance in the price of the land. A locality cannot be built up by farming on those lines as it can by the effort of farmers on moderately sized farms, for reasons that will be apparent.

The nature of the farming probably more than anything else should determine the size of the farms. Where the farmer grows only grain and does the work mainly himself, 160 acres is amply large for such a farm. Where he keeps live stock and must needs confine the grazing of them within the limits of his farm, he should have not less than 320 acres of grazing land in dry areas. Land in such areas does not produce so much pasture as in areas that are moist. When the farmer can control rough range pastures contiguous to his land or that form a part of it, he may need a section or two of rough land for each quarter section of arable land in his possession. On the latter he will grow the food that he needs for winter feeding. In yet other instances the farm may be all classed as rough land, and yet within it there may be enough arable valley land to enable the farmer to grow on these the winter food needed. These farms also should not contain less probably than one or two sections. Under irrigation the small farm unit is better than the large one for the average farmer.

While 320 acres may be called for where dry land farming is to produce the best maximum results in dry land, farming 80 acres would seem ample where irrigation is practised. Where the farmer who can use irrigating waters has more than 80 acres to care for, in a majority of instances, the evidences of a neglected tillage are more or less present. This holds true of lands that are farmed more or less even on the lines of live stock production.